HOW TO MAKE MONEY
Growing & Selling
MICROGREENS

AN INDOOR URBAN FARMING GUIDE

Brandon Keady

©2018 Brandon Keady
ALL RIGHTS RESERVED

ISBN-13: 978-1093294712

No part of this work may be reproduced, stored in a retrieval system, or transmitted in any form or by any means, electronic, mechanical, photocopying, recording, or otherwise, without written permission.

Although the author and publisher have made every effort to ensure that the information in this book was correct at press time, the author and publisher do not assume and hereby disclaim any liability to any party for any loss, damage, or disruption caused by errors or omissions, whether such errors or omissions result from negligence, accident, or any other cause.

Cover design by Zola Design Co.
zoladesignco.com

A NOTE FROM THE AUTHOR

Farming microgreens in your home can be fun, tasty and profitable. This book sets out to explain the basics of starting a microgreen business, from the planning stages, to growing, harvesting and selling. I have attempted to cover all the necessary information and make it easy to understand, without being too technical. You do not have to use every single method suggested; do what works best for you and your business. It is my hope that this book will save you many hours of research.

May you enjoy much success!

—Brandon Keady

TABLE OF CONTENTS

PART I - INTRODUCTION

Is the Microgreens Business for Me?	8
A Note About What Microgreens Are and Are Not	9
Why Microgreens	10
Super Foods	12
Super Taste	13
Can I Really Farm Indoors	14
Do I Have to Farm Indoors?	14
What If I Want to Start Small?	16
How Much Money Can I Make?	16
Some Things to Consider	17

PART II - BASIC SUPPLIES

What do I Need to Begin?	20
Trays	22
Shelving	23
Lighting	24
Ventilation	26
Growing Media	29
Hydroponics	30
Soil	30
Perlite and Vermiculite	32
Peat and Sphagnum	32
Coconut Coir	33
Rockwool	33
Fiber Grow Pads	34

TABLE OF CONTENTS CONT.

PART III - FARMING PRACTICES

Sowing	36
Soaking Seeds	37
Darkening Seeds	39
Weighing Down Seed	40
Watering	41
Temperature and Humidity	43
Nutrients	44
Buying Seeds	45
Harvesting	46
Packaging	48
Labeling	49
Refrigeration	50
First Crops	51
Troubleshooting	51

PART IV - BUSINESS PRACTICES

Before You Begin	56
Formulating a Business Plan	56
Choosing a Business Name	60
Registering Your Business	60
Keeping Minimal Overhead	61
Bookkeeping	62
Where to Find Customers	63
Fresh Sheets	63
Pitching Your Product	65
Developing Good Business Relationships	68
Marketing	69
Other Ways to Get Noticed	73
Branding	74
Final Words	76

PART I
INTRODUCTION

IS THE MICROGREENS BUSINESS FOR ME?

Microgreens farming is like any other business. It requires hard work, commitment, and a positive attitude. If you are reading this book, you are probably pretty curious and enthusiastic about the idea. Getting into business for yourself is a lot like beginning a new relationship. At first, it's very exciting, the possibilities seem endless, and it is difficult to see the downside. But, after some time has passed, the initial excitement fades and you realize that it takes effort. If you are still passionate about what you are doing, and are in it for the long haul, you won't mind putting in the hard work that it takes to be successful.

If you have never worked for yourself before, you might want to consider the following:

Do I have a genuine interest in microgreens farming or am I just trying to make a quick buck? This business is not a get-rich-quick scheme. It will require hard work and perhaps more than you are expecting. You may have to make sacrifices of your time in order to keep your business running.

Am I willing to take a risk? The great thing about microgreens farming is that compared to other businesses, the initial capital requirements are relatively small. But it is still an investment, and you should be aware that there is a risk associated. With careful planning, however, you should be able to turn a profit.

Even with careful calculations, things will happen that you cannot account for. Sometimes these problems require a bit of creative thinking to solve.

What if I don't have any gardening or farming experience? While you do not absolutely have to have any experience to start, it does help. For those who have never gardened, it would be wise to grow just enough for yourself and your family at first. Start off small before you decide to quit your job and launch into a full-time farming operation. Then work your way into a part-time or moonlighting operation before making it your sole source of income. This will give you an idea of the process, and to make sure it is right for you. There is potential to make thousands of dollars, even doing this business twenty hours a week.

A NOTE ABOUT WHAT MICROGREENS ARE AND ARE NOT

This book uses the term microgreens to mean the young shoots of any edible plant being harvested for consumption at a stage of development in which the plant has taken root and has grown its first true set of leaves (the third and fourth leaves).

Microgreens are not as fully developed as baby greens, which have been grown longer and are larger and leafier. Baby greens (usually lettuces) are cultivated to be harvested while young and tender, then left to regrow several more times, for multiple harvests. It should be noted that some microgreens can also be harvested and left to keep growing.

Sprouts, on the other hand, are plants that have just emerged from the seed and have not taken root or formed leaves. They are mostly stem. Needing no outside sources of light or nutrients, they are often grown in water. They may still have their seed hulls attached. Microgreens are grown for longer and often have a stronger flavor.

As this book refers to the farming of microgreens, the information contained within is not necessarily useful for growing sprouts or for baby greens. I do however encourage anyone who wishes to incorporate either into their business model to do so.

WHY MICROGREENS?

Microgreens are in high demand and quickly becoming one of the most popular food trends in North America. They are a fast-growing, high-value crop that can be farmed indoors, even in your own home. Lauded for their superb antioxidant and nutritional value, these miniature carpet crops have become valuable staples among chefs, juice bars and the health conscious. Although they were once used chiefly as garnishes in gourmet cuisine, they have entered mainstream dining experiences.

Microgreens are relatively easy to grow, require little space, do well under artificial lighting and cycle very quickly. Crops can be farmed all year round. Unlike other foods, they can be grown with or without soil, vertically in small spaces via shelving and be planted en masse with little spacing. You do not need a degree in agriculture, acreage or expensive farming equipment to begin. Neither must you quit your day job or move to the country to farm microgreens, which makes this business an ideal, low-risk and low-investment enterprise for beginners.

Compared to other agricultural industries, the cost of starting a microgreen farm in your home requires minimal capital. It is a business model that can be easily scaled to a size that fits your space and to the number of hours that you are able to invest. You can start a microgreen farm in your own home in your basement, spare room or garage. It doesn't require a lot of knowledge or special skill. It can be done on a small scale, at any time.

Not all forms of produce can be transported over long distances without deterioration. Microgreens, being fragile and prone to damage and wilting, fall into this category. Harvesting plants at the seedling stage cause them to be very delicate. This is why you may have seen them being sold unharvested and still in the substrate they were grown in for the consumer to pick. Once clipped, most have a shelf life of approximately one week, assuming they are harvested and stored correctly.

The problem this creates for businesses is solved by using fresh, locally sourced plants. This is where you come in. You will provide the solution by growing and delivering your crops to businesses within your own community. Microgreens are picked at the peak of perfection. They are not easily shipped, nor can they sit around on store shelves long. These are reasons why these tiny crops are valuable and garner top dollar.

SUPERFOODS

Pound for pound, microgreens pack a higher nutritional punch than their full-grown counterparts. Because they are harvested in the early stages of their development, they contain most of the nutrients still needed to form a mature plant, at a fraction of the size. It is estimated that some microgreens are up to 40 times more nutritionally dense than those grown to full maturity.

Commercial varieties that have been tested were high in the following vitamins:

- K: good for bones and blood clotting

- C: an antioxidant, boosts the immune system, prevents colds, promotes mental well being, and when taken with iron, increases its absorption

- E: also contains antioxidant properties, aids in good eyesight and helps to maintain healthy bones and skin as well as:

- Beta-carotene: boosts the immune system and promotes healthy vision and skin

- Lutein: known as the "eye vitamin", it may prevent several cancers, diabetes, and heart disease

- many other nutrients.

It is easy to see why microgreens are in high demand among those who are concerned with healthy eating and preventative nutrition.

Nutritive value is one of the key reasons people love to eat microgreens. It is a good selling feature for customers looking to consume the product themselves (something to keep in mind if you sell to individuals or at farmers' markets).

SUPER TASTE

Young and fresh and chock full of nutrients, microgreens pack a flavorful punch! Generally speaking, most microgreens take on a similar flavor to their adult counterparts. Chive microgreens have a chive flavor, radishes, a radish flavor, and so on. Some taste like more pungent versions of their mature selves, and some like milder versions.

It is always advisable to know the products you are selling. Before entering into the world of selling microgreens, I suggest you grow a variety for yourself and your family to sample. Try them in shakes and smoothies, in salads, on your burgers, as a garnish. Try ones you normally wouldn't and find out what is great about them. Always know your product better than your buyers. Be ready with suggestions to suit their needs. Your insight and confidence in your crops will procure more sales.

CAN I REALLY FARM INDOORS?

The short answer to this is yes! The amount of space you are able to utilize for farming will have an impact, of course, on the amount of money you are able to make, but even small spaces used effectively can produce big profits! It is possible to earn thousands of dollars per month in only a few hundred square feet of space.

Entrepreneurs all over North America are having great success farming out of basements and spare bedrooms. A garage, shed or greenhouse are other good options.

DO I HAVE TO FARM INDOORS?

No, but you probably should, at least to start. Growing indoors allows you to maintain nearly complete control over your crops. It is true that traditional farming is something long done outside, under the sun - arguably the best light source for growing plants. Growing microgreens outdoors, however, has some serious disadvantages. You will have to contend with the weather - too much rain, too little rain, or watering the crops yourself. The sun cannot be counted on to always be out and shining. When you grow indoors, you can control the amount of light and water your plants receive.

Growing outdoors also means you may have to deal with pests like bugs, rodents and other animals looking for a tasty snack, or the neighbor's cat digging around in the dirt. And even worse, animals and bugs will leave bacteria and contaminants on your plants and in your soil that could be potentially very harmful when eaten.

There are laws regarding what is and is not considered organic. If growing organically is of importance to you, determining whether your soil will produce organic crops is a must.

Weeding is an ongoing effort in any garden which may require many hours of labor. And because you will be planting your microgreens with no space between them, and because they will not be growing to maturity, you will not be able to mulch around them as you would for other garden plants. Also, seedlings are sometimes difficult to tell apart from weeds.

For these reasons, much of the farming aspects in this book will pertain to indoor microgreens growing. Inside is the first and best choice for many growers.

WHAT IF I WANT TO START SMALL?

If you are just starting out or testing the process for yourself before launching into a full-scale operation, you might begin with a spare closet or small grow tent. Grow tents are fabric tents that are lined with reflective silver colored mylar. They have plastic or metal framing and zipper shut so that light cannot escape. They are available in a wide variety of sizes and some even come in kits with everything you need to assemble them. You can find grow tents by searching out retailers near you or online (they are readily available at places like eBay and Amazon). Setting up a grow tent is not too difficult and might save your closet walls from moisture problems.

HOW MUCH MONEY CAN I MAKE?

Are you looking to supplement your income? Maybe make a few extra dollars or fun money? Perhaps you want to pay off your student loans, or payout your mortgage. This business is scalable to any of those things and bigger.

A willingness to work hard is the most important factor for success. Growing microgreens is not a 'get rich quick' scheme. It's a real business that requires industriousness. If you put in the effort to achieve your goals, you will be successful. How successful? Well... how hard are you willing to work?

If you were hoping for a straight-up answer in dollars and cents, I can tell you that as of the printing of this book, the current return for microgreens runs roughly between $20 - $50 per pound. Much of this difference is due to the variety of microgreens sold. A planted tray sells for about $20 in many places, although, depending on your location and competition, that number will fluctuate. It also depends on the species of plant and the quality of your product.

SOME THINGS TO CONSIDER

Oftentimes when people decide to start a business for themselves, they head into it with grand ideas and visions of their business solving all of their problems. Now, there is nothing wrong with having big dreams. But in order to succeed, you need goals. A goal is a dream that has both a plan and a deadline.

Starting a new business without a plan or a timeline is a recipe for disaster.

So before you read any further, I want you to think about what your goals are, both for the short term and long term. Formulate an idea of what you want to achieve. Keep that in mind as you read through the next sections of the book and think about how this information can help you achieve them.

Some things to ask yourself:

- How can I best utilize the space I have available?

- How much money do I need to invest?

- Do I have the time I need to devote to the scale of operation I have chosen?

- How can I be competitive in my area?

- At the end of the book, you will find a section with tips for putting together a business plan.

PART II
BASIC SUPPLIES

WHAT DO I NEED TO BEGIN?

In this section we will look into the necessary space, equipment and supplies you will need to begin. This book provides all the information you need to start farming microgreens. Nevertheless, you may wish to research or experiment with products to find what best suits your needs and business model.

In order to maximize profits and for optimal efficiency, a microgreens farm should be sectioned into different areas, each with its own purpose. The first, and most obvious is an area where your plants can grow and thrive under proper lighting. It should have an adequate water supply nearby. A utility sink with a hose is perfect for this.

Next, you will need a table or place to do your planting. If you are just starting off small or on a very tight budget, you might consider using your kitchen counter or dining table. If this seems impractical, a fold-up utility table will do. They can be stored and out of the way when you are not using them. You might also consider a secondhand table or two sawhorses with a sheet of plywood. You will want to use a washable plastic cloth over the latter for cleanliness.

A third area you will want to consider, though it is not absolutely necessary is a place to put your germinating seeds before they have sprouted, and before they require a light source. Microgreen trays at this stage can be stacked tightly together in relatively tiny spaces. Utilizing this third area of operation enables you to have a larger, steadier supply of microgreens and will increase your production.

You will also need a space to keep all of your supplies, from trays to watering devices, extra light bulbs, scoops, your growing media, seeds, cleaning supplies and all the other things that you will use on a regular basis.

Lastly, you will want a place to keep the business end of things - your books, receipts, etc. This may be a desk drawer or shelf. An organized computer folder, dedicated to your business will also be handy, plus any programs (spreadsheets, accounting software, etc.) you may want to use to help you keep track of the financial aspects of being self-employed.

TRAYS / FLATS

Unlike traditional farming which is done in fields over acres of land, microgreens are a tiny crop most often grown in shallow plastic trays. A wide variety of trays are available on the market to use. Some are marketed specifically for growing microgreens, while others are labeled more generically as growing, drip, seedling or propagation trays. They come in different thicknesses of plastic, sizes, and depths. Some have drainage holes and some do not.

A typical tray will be 18-20 inches long, 10-12 inches wide, and 1-2 inches deep. It should not be divided into small sections such as the trays that have tiny 1X1 inch divisions for germinating other kinds of plants. This type of tray is meant for plants that will be removed and transplanted into a larger container or garden.

Starting off with good equipment will save you a lot of headaches. Get the best you can afford. A good microgreens tray will be at least an inch deep and remain fairly rigid when filled with damp soil. It is also possible to stack two trays together for extra support. Trays that have drainage holes will expel excess water easily, but could possibly cause the soil to dry out quicker. Microgreens are fragile and germinating plants require moist soil. Deeper trays which hold more soil will retain moisture for longer periods of time because water will not evaporate from them as quickly. However, the weight of the added soil could make the trays more unstable and prone to cracking. Microgreens require very little growing medium as they are not being grown to maturity. Extra trays and soil are expenses you do not need.

Disease and mold can be passed between crops. It is important to keep trays clean by washing and sterilizing them between crops. Use hydrogen peroxide or bleach to sterilize them. It is important to rinse any bleach out.

SHELVING

The size of your grow room will determine the number of crops you can comfortably grow at one time. In order to have a large number of plants in a minimal space, shelving is a good choice as it allows for vertical growing. It is possible to stack plants on shelving, up to 4 shelves, with proper lighting, within the confines of a standard 8-foot high ceiling. This setup, however, does not enable you to easily see all your crops at a glance.

Another option, if you have the room, is to lay your trays out on one flat surface, table or counter high. This is probably easier, as they will be easily visible, and easy to water and care for.

There are many options for shelving and tables. The easiest setup may be to purchase inexpensive steel wire shelving kits which can be put together with a few simple tools. And folding utility tables are an easy choice for table setups. Shelving, however, may be preferable as there is framework there to affix lights to.

LIGHTING

Sunlight provides plants with energy to make food. This process is called photosynthesis. A green pigment in plants known as chlorophyll allows plants to absorb light which is then converted into glucose, a form of sugar. Providing your microgreens with adequate lighting will result in faster-growing, more robust crops.

Sunlight is free and available anywhere. If you have a greenhouse, this may be the easiest way to go. Even so, sunlight is not always consistent, and this may affect your ability to provide a steady source of crops on a schedule.

It should be noted that many microgreen plants do not require full sunlight as it can damage young leaves. This is especially true of lettuces which grow better in partial sun. If you are planting outdoors or in sun, these microgreens will benefit from indirect light (very bright daylight without the full strength of the sun). Most plants need at least 8 hours of indirect sunlight per day. If you are struggling to get enough light, try creating a reflector. Reflectors are used to bounce light back onto the plants, the same way they are used to help brighten subjects in portrait photography. You can make one out of white foam board or cardboard. Alternatively, reflectors could be covered with Mylar.

If you are setting up your farm indoors, you will probably need to provide an artificial light source for your microgreens. The sun cannot be replicated, and is what is best for growing plants under most circumstances. That said, microgreens do not require much in the way of lighting, as they are not being raised to maturity.

In choosing a lighting system, there are several things to consider. Many different kinds of lights can be used to grow microgreens. Lighting comes in a wide range of wattages/lumens and intensities. In terms of color spectrum, microgreens tend to grow well under a full spectrum light, which mimics sunlight. This is not to be confused with 'daylight' bulbs which often only give the appearance of daylight but are not full spectrum.

Without getting too technical, let us examine artificial light choices for growing microgreens:

- CFL (compact fluorescent lamps/lights): Fluorescent lighting is a popular choice among microgreens growers. Flourescent lights are affordable, last a long time, and produce minimal heat. The light created by these bulbs is dispersed in all directions.

- LED (light-emitting diode) lights: While a more expensive initial investment, LED lights last a very long time and are much cheaper to use long term. They emit virtually no heat. The light created by these bulbs is directional.

- Incandescent and Halogen bulbs: These are NOT recommended for growing microgreens for the following reasons: Although cheap, they do not last very long and must to be replaced often. Plus they use a lot of energy which will translate to dollars on your electric bill. Incandescents also generate heat, which can burn plants.

No matter what kind of lighting you choose, you may need to

experiment with your setup, particularly the proximity of your light source to your trays. Weaker lights can be placed closer to growing plants to achieve better growth. It is best if your lighting system is easily adjustable. I suggest light housing hung from chains that can be adjusted up and down on chain links. A good example of this is the typical fluorescent shop light. Shop lights are commonly employed by microgreen growers. Alternatively one could set up a system where the plants are moved and the lights remain stationary, perhaps by platforms.

Another thing that you can do to ensure adequate lighting is to reflect the light from your bulbs off the walls of the room you are growing in. Paint the walls white, or hand white plastic sheeting over them, or even better, cover them with Mylar, a silver colored, highly reflective material as these will reflect light best.

VENTILATION

Before launching into a full-scale microgreens business from your home, take some time to carefully think about the room you will be farming in. The byproduct of growing a large number of plants indoors is an increase in temperature and humidity. These two elements create the perfect environment for bacteria and fungi to develop, which can detrimentally affect your crops. Even worse, this combination can cause mold and mildew in your home that could affect your home and your health. Mold can be dangerous and difficult to eradicate. Good planning will prevent hazards like this.

Microgreens require good ventilation for their development. Because they are grown tightly together, the amount of space around each plant is minimal. As they take in carbon dioxide to grow, they expel oxygen. Pockets of oxygen will form around overcrowded crops, creating less than ideal growing situations. By moving the air around, your plants will be healthier.

Air flow is also important for limiting humidity. Warm, stale humid air is an excellent environment for fungi and pathogens to grow. Proper ventilation will remove excess heat and humidity. The following information should be used as a base guide for what you may need. Please take into consideration that every setup is unique, from the size and shape of rooms and the amount of moisture in the air, to your climate, altitude, and the seasons.

The first step in creating good ventilation, and perhaps the easiest thing you can do is to install a motorized, fan or fans in your grow room to push the air around and create an air current. Fans that generate minimal heat are preferable. Aim to create a light breeze that gently moves the tops of your microgreens around.

Larger operations will require more fans and/or possibly exhaust fans. Exhaust fans draw the warm, humid air and expel it outside. (Think of a bathroom exhaust fan that you use when you have a shower. Hot, humid air rises, gets sucked out by the fan, usually into a duct where it is expelled through a vent on the roof. Cooler air flows under doors and other openings, replacing the warm air.) This creates a constant supply of fresh, circulating air. Exhaust fans will be rated CFM (cubic feet per minute), which is how quickly they are able to exchange the air. Aim for a system that can exchange the air three times per minute. To figure out the size you need, determine how many cubic feet your room is and multiply by three (to find the CFM first measure the dimensions of your room in feet, then multiply the length X width X height X 3).

Air conditioners are also able to remove moisture and create air flow. An air conditioner may or may not be necessary depending on the climate, temperature and time of year. The following is only a guideline.

Air conditioners are determined in BTU (British Thermal Units). To determine the BTU rating you need for your space, first, find the square footage of your room (to find the square footage measure the dimensions of your room in feet, then multiply the length X width). On the next page you will find a chart of how many BTUs are required per square footage of space.

While these numbers reflect your needs based on room size alone, they do not take into account heat from lighting and ballasts and any appliances, which will increase cooling requirements.

Lastly, if high humidity levels are a problem for your crops or your home, a dehumidifier will remove moisture from the air. This is a great solution for smaller setups that only use fans to move air around. However, if you are circulating large amounts of air into and out of the grow room by way of exhaust fans, the positive effects may be negligible. The key to lowering humidity in this situation is to have dryer air coming into the room than is going out.

AREA OF ROOM	BTUS REQUIRED
100 - 150	5000
150 - 250	6000
250 - 300	7000
300 - 350	8000
350 - 400	9000
400 - 450	10,000
450 - 550	12,000
550 - 700	14,000
700 - 1000	18,000
1000 - 1200	21,000
1200 - 1400	23,000
1400 - 1500	24,000
1500 - 2000	30,000
2000 - 2500	34,000

GROWING MEDIA

Microgreens are best grown in media with a pH (acidity level) of around 6-6.5. You can test the pH level of the medium you are planting in using a kit found in most garden centers and hardware stores. Microgreens can be grown in a variety of media. Due to their small size and the nutrition initially provided by the seed during germination, they are not heavy feeders and do not require as many nutrients as plants being grown to maturity.

HYDROPONICS

Hydroponics is the term used for growing plants without using soil. Many microgreens can be grown without soil. A seed contains everything a new plant needs to begin germinating except water. Adding water to a seed starts the germination process. The seed provides all the nutrients the plant needs for this initial phase of growth.

Plants emerge from their seed pods with a set of 'leaves' called cotyledon. Growth after this stage requires sunlight and additional nutrients and water. While many microgreens can get to the next stage of development, where the first new, true leaves are formed, without the addition of nutrients, it is often noted that plants do better with them. Hydroponic systems utilize additional nutrients to grow healthy plants. You can perform your own experiments to test how plants fare with and without the use of nutrients in a hydroponic system. Other things you should consider are price, availability, and ease of use.

SOIL

Potting soil is often the winning choice for microgreens farming as it is natural, cheap and can be found anywhere. It is easy to work with and if you've ever had a garden or even a houseplant, you already have experience using it. Plus, it's hard to beat what nature intended. The one downside to using soil is that it can be messy. Your harvests may need to be washed to remove traces of dirt.

Soil is the most practical medium to use for plants that have larger seeds like sunflowers and peas. An important factor when choosing soil is texture. You want a soil that is very finely textured and spreads uniformly. Avoid any soil with large chunks as your plants will not be able to send up shoots through them.

Topsoil can and should be sterilized if you have dug it yourself or it has been outdoors and you plan to grow microgreens indoors. Otherwise you risk bringing in all kinds of contamination including bugs and pathogens to your farm. To sterilize soil, put it in a large baking dish or roast pan and bake in your oven at 250 degrees for one hour.

Topsoil also compacts and hardens easily. The addition of perlite, vermiculite, peat/sphagnum, and coconut coir will improve topsoil's texture. These amendments, particularly the latter three, will also help to retain moisture. Compost is another fine addition that aids in water retention and adds nutrients. As with soil texture, ensure the particulates are very small.

PERLITE & VERMICULITE

These two natural substances are added to potting soil for water retention and aeration. Microgreens can even be sown directly into these two mediums although it is probably not very cost effective to do so. These substances are slightly more alkaline (not acidic enough) than what is optimal for microgreens. Amendments may be necessary to achieve the correct pH.

PEAT & SPHAGNUM

Peat and Sphagnum are two different parts of the same plant which grows in bogs. They are used for improving water retention and soil texture. Sphagnum is the green moss you see used in dried floral arrangements. It is pH neutral. Peat is not purely moss and may contain twigs and other organic matter. It is high in acid and tannins. If your media contains peat you may need to adjust the pH. While both are completely natural, they are nonrenewable resources.

COCONUT COIR

Although it looks and feels a bit like soil, this medium contains none. It comes from the fibrous husk of coconut shells. When used in growing it provides almost no nutritional value for plants. If you choose coconut coir you will basically be growing your microgreens in water - a hydroponic system. If you plan on letting plants develop past their first two leaves, or if you will be doing multiple harvests from one planting, this is probably not the best medium to use, without additional nutrients.

Coconut coir comes in blocks, bricks, and chips. After the condensed blocks and bricks are soaked in water, they expand to many times their size and are easily crumbled and spread out evenly.

* Coconut peat is similar in texture to peat moss and resembles ground up coconut coir.

Coconut coir is extremely absorbent. Plants are less likely to dry out with this medium. It can be used multiple times and does not break down easily. It is an environmentally friendly, renewable resource.

ROCKWOOL (STONEWOOL)

Rockwool is considered one of the best growing mediums for hydroponics because of its amazing ability to retain water. It was first used as insulation in the building industry and is made from heating chalk and rock to very high temperatures, then spinning and cooling it (a bit like cotton candy) into a soft, fibrous material. Rockwool is sterile and can be re-used but is purportedly alkaline, although some people disagree. Its acidity level should be tested to determine pH levels. Rockwool should be soaked in treated water to lower its pH before using. Regardless of its pH, rockwool should be soaked in water prior to use. It should also be noted that it can also be irritating to skin, eyes, and lungs.

FIBER GROW PADS

Fiber growing pads are often made out of natural substances like hemp, wood fiber or coconut coir. They resemble dense, fibrous mats. Some are cut to fit seed trays and some come on rolls that you cut to fit yourself. The best thing about grow pads is how easy they are to use. This makes them popular with microgreen farmers. They are not messy and can easily be dried out between crops and reused. They are more expensive, however than soil.

To use grow pads, you simply add water, spread your seeds over top and mist with water. Nutrients can be added to the water supply. Plants with small seeds root well in fiber grow best with grow pads. It should be noted that grow pads often require soaking or rinsing before use.

PART III
FARMING PRACTICES

SOWING

Microgreens generally grow best when spread 1/8 to 1/4 inch apart. This allows the plants to grow thickly yet still have some air flow between them.

Before you sow, be sure to measure your seeds. Measuring allows you to gauge exactly how many trays you will get per pound of seed, and make it easier to know when and how much to order. But some prefer to eyeball it, and perhaps this method will work well for you.

If you find it easier, press down the surface of your media to create a flat, even surface on which to spread the seeds. A flat board or tile with a handle works well for this, but you could simply use your hands. You might also want to water your media before putting down your seed, especially for tiny seeds which are easily moved about by spraying.

The easiest way to sow dry microgreen seeds is to sprinkle them over your growing medium, then spread with your hands. This method isn't perfect but is quick and works very well for many varieties. Very tiny seeds can be mixed with a small amount of soil for more even sowing. For seeds that have been soaked and are mucilaginous (jelly-like after soaking), drop in small amounts around your tray and spread them out by hand, as evenly as possible. After your seeds are down, another layer of media can be sprinkled over top, if necessary.

Another option for sowing seeds is to use a dibbler/seeder. A dibbler is a tool used to poke holes into the ground in which to sow your seeds. Dibblers come available in sheets that are able to create a large number of holes at the same time. This method creates the exact same spacing between seeds and even growth in your tray. Dibblers may be ideal for growing baby greens or plants to maturity, but are not a necessary piece of equipment for microgreens. This method is also time-consuming.

If your plants are sprouting through the soil with too much dirt clinging to them you may want to consider using a lighter, finer soil or medium for covering them, such as vermiculite. Alternatively, you could add a layer of paper towel over the seeds and moisten it. This allows for germinating plants to break through the surface more cleanly.

Some seeds, if not covered with soil, will need to be darkened in order to germinate properly. A layer of soil overtop or a weight placed on them will force them to root properly and create a stronger, more stable plant.

SOAKING SEEDS

The purpose of soaking seeds in water before planting is to soften the hard, outer hull of the seed so that it splits open more easily when germinating. Soaking seeds can reduce their germination time. But not all seeds need to be soaked before sowing, and it is more important for larger seeds with tougher exteriors. Peas are a good example. To soak seeds properly, measure the seeds into a container and add three times the amount of water. Be sure to use room temperature to warm water. Do not use hot water as it can cook the seed.

Soak according to the instructions on the seed packet. You can always experiment by soaking longer to determine whether it speeds up germination. It is natural for seeds to expand or change color a bit during this process. Discard seeds that float, as these are likely duds and will not sprout. If soaking for longer than 24 hours, drain and add fresh water, after a day has passed.

To drain seeds simply pour the water off by hand. But feel free to use a colander or mesh strainer if that is easier. Some seeds are very tiny. If the holes of your strainer are too large, pour off excess water and spread by hand, or place a coffee filter in your strainer.

Some seeds are mucilaginous; they form a gel sac around the seed

when exposed to water. This helps to keep the seed moist at all times. Do not remove the gel sac.

DARKENING SEEDS

Most of the time when growing microgreens you will be surface sowing - spreading the seeds directly over the top of the earth instead of burying them. In some instances, these crops will require darkness in order to grow. The layer of soil that would normally cover them would keep them in darkness during the germination process. Artificially darkening seeds allows for plants to germinate more cleanly, with less soil clinging to them.

Darkening your trays is as easy as taking an empty tray and placing it over top of the planted one like a dome, or upside down with a weight inside. Trays can also be placed in bags or dark rooms and closets. Planted trays can even be stacked. This works well if planting multiple trays of the same variety at once. It is also a good solution for plants that require a weight as well as darkness. Cover the top tray with an empty, or weighted one. If you use a bag, be sure to leave the end open for air circulation.

After 2-3 days germinating, the plants should be ready to be uncovered. You will notice that they are yellow and pale. This is because light is needed for photosynthesis and to happen. After a day of exposure, you should see a noticeable difference in color.

WEIGHING DOWN SEEDS

Some microgreen seeds sown directly onto the surface of the soil will require a weight. In nature, germinating plants must push up through the soil in order to find sunlight. The weight of the soil forces the plants to root properly. This makes for sturdier plants that will not topple over easily. In growing microgreens, we are not really following what happens in nature. Providing weights is an artificial means of replicating the layer of soil that would naturally cover them in nature.

Why not cover them with a layer of soil? Well, you can if you prefer. It is often cheaper and easier not to. Having a weight you can lift, however off lets you keep an eye on how your microgreens are growing. And as the plants germinate they will have less dirt clinging to them as they grow.

The easiest way to weigh down seeds is to stack your trays, one on top of the other. You can add a planted tray to the top that either does not require a weight, or one filled with soil, tiles or boards for the extra pressure. Be careful not to add too much though as this will compact the earth and make it more difficult for plants to grow.

As soon as the plants have germinated (approximately 2-3 days), and the weighted ones have driven their root down into the soil, the lid/weights can be removed and crops can be placed under lights.

WATERING

All plants require moisture to survive and thrive. Water is the most important aspect of germination. The best water to use for watering your plants is unchlorinated, filtered water. Chlorine is the chemical we associate with the smell of swimming pools. A small amount is added to municipal tap water to kill bacteria. This makes it safe for human consumption. But chlorine, even in tiny amounts can affect plants' growth. An easy way to remove chlorine from tap water is to let the water stand for at least 24 hours so that the chlorine has time to evaporate. If you have ever had an aquarium, you may be familiar with doing this when performing water changes.

The pH Scale

Battery	Stomach Acid	Lemon	Soda	Tomato	Coffee	Milk	Water	Blood	Egg White	Stomach Tablets	Ammonia Solution	Soap	Bleach	Drain Cleaner
0	1	2	3	4	5	6	7	8	9	10	11	12	13	14

Acidic — Neutral — Alkaline

Most microgreens thrive at a pH level of around 6 - 6.5. Using water at this acidity level is the best choice for watering your crops, whether you are growing in soil or hydroponically. In terms of acidity/alkalinity, a lower number is more acidic, and a higher number is more alkaline. Water can be easily tested with litmus strips which you can purchase at any garden or hardware store.

- LOW PH: If your water is too acidic (this is rare), a very small amount of baking soda can be added to it. Re-test to make sure you have achieved a correct pH level.

- HIGH PH: If your water is too alkaline, a small amount of vinegar or lemon juice can be added to it. Re-test to make sure you have achieved a correct pH level.

Before watering, check trays to see if your growing media is dry. Instead of putting a finger into the soil which might damage fragile plants, it is better to lift the edge of the tray and feel for overall weight. The lighter the tray, the less water it contains. Soil pulling away from the edge of the tray is also an indication of dryness. This method may take a little getting used to but is better than disturbing your crops. Not all varieties will have the same watering needs.

Underwatering is an easy way to water delicate microgreens. This method requires perforated growing trays with drip trays below them. Cafeteria trays are a good alternative to drip trays and do not need to be replaced. Add water to the drip tray instead of directly to the seeds/plants. It will penetrate the perforations of the growing tray above, into the soil, and the water will be fed directly to the plants' roots. The benefit of watering this way is that the microgreens do not get wet. Wet plants are more prone to bacteria and fungi. Also, smaller seeds will not pool or get moved around.

If your farm is located in a greenhouse or outdoors, a hose on a very gentle mist or spray will also suffice to water your plants. Alternatively, a mister, like the kind in grocery store produce aisles, can be used and even be put on a timer.

Even a spraying bottle with a finger trigger will work, although if you have many trays to water your hand may get tired. To save your fingers, a homemade sprayer can be made using a disposable water/soda bottle, with a few tiny holes drilled in the cap. Fill with water and turn upside down.

TEMPERATURE AND HUMIDITY

If you are growing microgreens indoors in your own home, temperature should not be much of a growing factor. What is comfortable for humans is usually comfortable for plants. Most plants will grow in temperatures ranging from 50F (10C) to 85F (29C) but do best in a warm room with at least 50% humidity and good air flow.

Microgreens that are exposed to some humidity are crisper than those grown in dry air. If the air in your growing environment is too dry, add humidity to the air around your microgreens by underwatering. Use trays with drainage holes and putting a larger tray, such as a cafeteria tray below. As previously mentioned, the cafeteria tray can be filled with water which will water your plants from the bottom up. As it evaporates, it will also add humidity to the air around your plants. Or you may wish to add a humidifier to the grow room. Daily misting with water is another good way of maintaining humidity. Plants that are staying wet for a long time or that are exposed to excess humidity are prone to fungal infections and disease. It is important to strike a good balance. Too much humidity is more of a cause for concern than too little.

Set up a fan(s) around your plants to keep the air flowing between them if your humidity levels are too high. The fans will have the added benefit of moving the microgreens around slightly, and result in stronger, healthier plants. But don't overdo it. All you need is a gentle, oscillating breeze. Too much air pressure in one direction will cause them to lean. More about this under the ventilation section.

NUTRIENTS

If you are growing microgreens in soil, you should not need to add any extra nutrients in the form of fertilizer to your soil. Plants do not need anything but water to germinate and emerge from their seed hulls. They have everything they need to form their first two embryonic leaves called cotyledon. As new, true leaves form however, they will require more than just water to be healthy.

If you have purchased potting soil as your growing media, it should contain enough nutrients to grow microgreens. If you are reusing it repeatedly though, you may want to turn under the remaining stalks and roots of the harvested crop as this will add nutrition back into the soil. You can also add nutrient-rich compost. Be aware that pathogens and fungi can be spread from one crop to the next by reusing your media.

As many people are very conscious about what is in their food and how it is grown, you may want to steer clear of non-organic fertilizers. There are many alternatives to choose from including compost, humus, mineral powders and many other commercial supplements in powder and liquid forms.

If you are using a strictly hydroponics growing method, you will want to add nutrients as this will improve the growth and overall quality of your product.

BUYING SEEDS

Because their foliage and roots will never have the chance to fully develop before harvest, microgreens are planted tightly together. This means you will need to purchase seed in larger quantities than the average person who is planting a backyard garden and buying their seed in tiny paper packets. One of those tiny packets will not cover a standard growing tray and will cost far too much. And depending on where you live, you may not necessarilybe able to find what you need down at your local department store. Instead, you want to find a reputable company with good quality seed that has excellent germination rates and sells in bulk. Microgreens seeds can be purchased by the pound and in large quantities.

Use a scale to weigh out your seed to give consistent results for each tray. The larger the seed is, the more is required by weight to sow a tray. Determine how many trays worth of seed you can get per pound. For example, if two ounces of seed covers a tray, you will be able to get 8 trays per pound of seed. This is extremely helpful in purchasing and for calculating profits.

The type of seed you choose will depend on what you are planting and your overall goals. Will your farm be 100% organic? Do you plan on buying heirloom or hard to find varieties? Whatever you choose, be sure to purchase quality seeds with germination rates 80% and higher (the higher the better). If you search online you should be able to find seed companies that specialize in microgreens.

HARVESTING

Most microgreens are ready to be harvested about 2-3 weeks after planting when they have their first set of true leaves and possibly second. If grown past this stage, many varieties will develop a bitter taste. The following is a step by step guide on how to harvest your crops:

- Water your microgreens 10-12 hours before cutting. You never want to harvest wet microgreens as excess moisture will result in wilting when packaged.

- When harvesting, make sure your hands and tools are extremely clean. You may want to use a disinfectant such as hydrogen peroxide (this is probably more important if you choose not to wash your greens before packaging).

- Before harvesting brush the tops of the plants with your hand to loosen and remove any seed hulls that may still be attached.

- Cut the shoots right above the soil line using sharp scissors or shears that have been clean and sterilized. If you are starting a large scale operation you might consider a harvesting machine which is efficient and consistent. For species that can be regrown, cut above the first set of leaves.

- Washing crops is not always necessary. If you choose to do so, wash the shoots in a clean sink or bucket of cool, fresh water to loosen dirt and remaining hulls. Avoid using strong hoses or sprayers. Do not use hot water as this will wilt your microgreens. And do not soak for any longer than necessary. As you pick them out of the water, push microgreens down first to separate seed hulls. Use a hand strainer to easily clean up any loose bits leftover in your sink.

- Dry your microgreens as quickly as possible. A salad spinner is a great way to remove most of the water and often the seed hulls as well. Afterward place them on a clean towel or paper towel to finish drying. Alternatively, you can spread wet microgreens over a drying rack (window screen or small mesh works well) in a thin layer to air dry. Use a fan to speed up the drying process.

- Using a scale, weigh your harvested microgreens place them into the package of your choice. Add your label. Labels can be used to help seal the bags or container shut.
- Refrigerate immediately.

PACKAGING

Harvested microgreens should be packaged to look appealing and professional. There are three options when packaging:

- Transparent Clamshell Containers: This is the best option, especially for grocery stores or when presenting in displays. Most people prefer containers to bags as they can see what they are getting. Clamshells are stackable, made with recyclable material, provide good aeration and prevent delicate microgreens from being crushed. Because microgreens are a specialty food item, most people will be willing to pay a few cents extra for the premium packaging. Make sure any plastic packaging you use contains ventilation air holes or your plants will deteriorate quickly.

- Bags: You may choose to use bags because they are more readily available or cheaper than clamshell packaging. For displaying, choose bags that have folds and stay upright such as poly lined, tin tie, paper bags with clear windows. It is important that bags have extra air around the microgreens to prevent moisture buildup and bruising. Plastic or poly bags are very cheap, but not recommended except for customers who will be consuming them very quickly.

- Reusable Bins/Plastic Containers: This is a good option for restaurant deliveries because they buy in bulk. They allow you to take old containers with you at delivery time, wash and reuse them.

Quality packaging lands customers and sets you apart from your competition. Put yourself in the customer's shoes... would you buy your product, or your competitor's?

LABELING

Part of presenting an attractive package is labeling. While restaurant chefs may not care so much, this is important if you are selling to individuals and grocery stores. Your product should appear professionally packaged. The easiest way to produce a label is to purchase pre-cut stickers in sheets such as shipping labels, that you can print as you need with a computer and printer.

Every label should have:

- Your business name (and logo).

- Your contact information (if you want direct sales).

- Your website address (if you have one).

- The product contained in the package.

- A date (example: harvested on, best before, etc.).

- Any other aspects of product branding.

- The words, "Delivery Available" (if you deliver to customers).

REFRIGERATION

Refrigeration is paramount to keeping harvested microgreens fresh. As soon as microgreens are packaged, they should immediately be put into a fridge. This is especially important in warm weather.

It is also a good idea not to transport them in a hot vehicle without a cooler. The most important thing to remember is to never refrigerate microgreens while they are wet. They should be as dry as possible so that they do not perish. Being delicate plants, this can happen very quickly.

FIRST CROPS

Record everything accurately! You will want to keep track of each seed that you plant:

- the amount of seed per tray
- media and methods used in planting
- length of time they took to germinate
- information about lighting
- watering - how much and how often
- the number of days until ready to harvest

This information will make it easy to plan future crops and for proper timing of harvests for customers. You should experiment with the process to see what works best for you.

Start small. This will let you get comfortable and will help you get accustomed to individual species better. Begin with no more than 4 different kinds of plants. Try ones that are known to be easy to grow first, such as radishes, broccoli, mustard, cress, arugula, and lettuce.

TROUBLESHOOTING

PROBLEM:
My microgreens still have seed hulls attached to them.

SOLUTIONS:
1. Make sure you are soaking your seeds long enough to soften the hulls before planting.
2. Use a salad spinner after harvesting to remove hulls (works for some species, usually bigger seeded ones).
3. Try a different, coarser medium.
4. Experiment with different seeds or seed company.

* * * * *

PROBLEM:
I am experiencing mold or fungus.

SOLUTIONS:
1. Make sure your supplies are clean and not harboring bacteria (hydrogen peroxide is perfect for sterilizing).
2. Do not overwater. Too much water will cause your roots to rot.
3. Your seedlings may be planted too close together. Plants need a certain amount of space around them to 'breathe'. When spaced too close together, too much humidity forms causing moisture to get trapped between plants. Plant fewer seeds per tray.
4. Use a fan to move the air around and through plants.
5. If humidity is an issue use a dehumidifier or exhaust fan to take moisture out of the air.
6. If none of the previous were the cause, it could be the seed itself. Sometimes seeds have fungal problems which may or may not be noticeable. The way to solve this problem is to sterilize your seeds. Soak seeds for 10-15 minutes in a quart of water to which you have added a tablespoon of white or cider vinegar and a

tablespoon of food grade hydrogen peroxide. Rinse very well.
7. Try a different seed company if fungal issues still persist.

* * * * *

PROBLEM:
My microgreens are yellow or pale.

SOLUTIONS:
1. If you have just taken the weight or cover off of a germinating tray of plants they will likely be yellow because they have not been exposed to any light. This is common. Place under adequate lighting and crops will 'green up' after a day or so.
2. Pale plants past this exposure to light stage may not be receiving enough light. Adjust your plants closer to the light source or use stronger lights.

* * * * *

PROBLEM:
My seedlings are falling over.

SOLUTIONS:
1. This could be caused by inadequate lighting. When plants do not receive enough light they get 'leggy'. They grow taller than they should, trying to reach the light source, lose vigor and rigidity and fall over. Use stronger lighting.
2. Another reason could be that root structures are not strong enough to support them. Weigh down and/or darken trays

before exposing to sunlight.

* * * * *

PROBLEM:
My plants are not germinating or are germinating too slowly.

SOLUTIONS:
1. Most microgreens begin to germinate within 5 days. Some species take longer. First, check seed packets to ensure you have correct germination times.
2. Give more water and keep media moist.
3. Seeds may be old or have perished. Try a new, fresh package of seed. Check dates on packaging.

* * * * *

PROBLEM:
My microgreens look sparse.

SOLUTIONS:
1. You may not have planted enough seed in a tray to adequately cover it. A properly planted tray will look like a thick, lush carpet at the time of harvest.
2. Not all of your seeds may have sprouted. This could be due to a number of variables such as adequate water, the freshness of the seeds, mold or rot, soil compaction or large particles in media, not darkening or weighing down seeds, etc.

* * * * *

PART IV
BUSINESS PRACTICES

BEFORE YOU BEGIN

Before starting any new business venture it is wise to do research in your country and state/province to know which laws and regulations you must abide by. Here are a few things to take into consideration:

- Do I need a business license?
- Do I need insurance?
- How do I register a business name? Can I use my own name?
- Do I need any special permits?
- What health and safety regulations must I follow?

Do not be intimidated. Most business owners are a lot like you. They are just regular people that had a lot to learn in the beginning, too. The successful ones are the ones that know their product and their market well. They researched and calculated risks before diving in. They have persevered through and learned from their mistakes. And they were not afraid to work hard.

FORMULATING A BUSINESS PLAN

This kind of business plan is for your own purpose. It is not to be confused with a formal business plan that you would bring to a bank. The purpose of this plan is to outline your goals and the means by which you will achieve them.

It is always exciting to be in the development stages of a new project. But before jumping right in...

- Research! Do your homework. Reading this book is a good first step, but it cannot tell you everything you will need to know, for example, your local market needs. Perhaps, after much research, you will discover that this is not the business for you, after all. And that is okay. It is far better to know now, than after a failure or financial loss.

- Determine who your customers are. Will you be selling to restaurants? Health food stores? Grocery stores? Farmers' markets? Are you filling a need in your community?

- What opportunity is there? Do you have competition? How will you make your product better and stand out more?
- Figure out the best way to utilize the space you will be working out of. Don't forget to take measurements.

- Determine how much capital you are willing to invest. Make this number firm, and never go over it.

- Finally, what do you need to earn in every month to make it all worth your while?

Add up your starting costs and first 6 months' worth of expenses. A half-year time frame is suggested because this is an adequate amount of time to establish your first clients, a routine and a regular stream of income.

LIST 1: List the essential equipment you need to start (trays, shelving, seeds, packaging, etc.). This list should only be the absolute necessities for growing and selling. Leave non-necessities for when you are turning a profit. Price out your list. Factor in taxes and shipping costs. Tack on an additional 25% for unforseen items, because even with good planning you will not be able to account for every possibility.

LIST 2: List your expenses for the first 6 months. This list should include vehicle expenses (gasoline), electrical, the cost to rent a table at a market X the number of times you will be using that table over the next 6 months. This is the cost of doing business for your first 6 months. Add 25% more to your total for a buffer.

LIST 3: This may difficult with no farming experience under your belt, but try to anticipate what your monthly expenses will be. The things in this list will be a combination of LIST 2 and things from LIST 1 that must be replenished from time to time (cracked and broken trays, more packaging materials, etc.)

Add up the numbers in LIST 1 and LIST 2 to estimate the amount of money you need to get your business up and running for the first six months. Do not neglect to add in the buffers. Having them ensures that you will be able to cover the cost of unexpected expenditures. You might even want to make this buffer bigger.

Check out the competition. Research what microgreens are selling for in your community. Do they look fresh? What varieties are they selling? How much are they selling for? Find out what you will need to be competitive.

Determine the amount of money you will need to earn per month to cover your monthly expenses from LIST 3 and make a profit. Is this amount an attainable with what you have to work with in terms of space and man hours? Do you need to make adjustments to your plan?

While sufficient planning is necessary for a successful start in business. It is important not to get stuck in the planning stages. At some point, you will have to jump in the water and get wet.

CHOOSING A BUSINESS NAME

You will want to choose a business name that is both memorable and reflects your product. Are you growing microgreens for the taste? Will you be selling to chefs? Or are you growing for health benefits? Your name should be easy to spell and remember. If you decide to operate under a business name, you will need to legally register it.

Though it may be easier to operate under your own name to avoid legal red tape and hassle, unless you have an established name or farming business in your area, a business name usually sounds more professional.

REGISTERING YOUR BUSINESS

In the United States, you should register your business name. This is necessary unless you are operating under your own, legitimate name (example: Michael Smith). Check with your state or county clerk's office for requirements. You may need to perform a search, publish a notice in the newspaper and pay a small filing fee.

Registering as a sole proprietorship is usually the cheapest and easiest way, but as the sole proprietor, you will be personally responsible for your company's finances. Registering as an LLC (limited liability company) has advantages such as protection from the debts and liabilities (creditors and lawsuits) of your business, but requires annual (and possibly other) fees. Sole proprietorship and LLCs may be taxed differently.

If you have a business partner embarking on this venture with you, you will want to look into forming a legal partnership agreement and structuring your business accordingly.

KEEPING MINIMAL OVERHEAD

If you plan to grow microgreens in your home, your overhead expenses will be substantially lower than if you lease a space from which to operate. The cost of paying monthly rent on top of the added electrical expenses for running fans and lighting will feel most significant in the beginning while you are building a customer base.

Set a budget and stick to it. Do not burn through your initial capital all at once. Avoid shiny object syndrome. Only invest in what is absolutely necessary. Do you really need to purchase shelves or can you make your own for cheaper? Do you really need a mister when you can water plants yourself? Can you find any supplies secondhand? Wait until you are showing profits before reinvesting. The quickest way to kill a business is to run out of money and the ability to produce anything at all.

BOOKKEEPING

Set up a bank account specifically for your business. Do not use your personal bank account for conducting business transactions. You need to be able to see your income and expenses, all in one place, at a glance.

Always have a clear picture of where your business is at financially. Keep track of every financial transaction. Accounting software will help you with the financial aspects of your business.

Create invoices for your customers, a simple summary of the goods and sum totals due to you.

Set up a ledger/software for money coming in and money going out. Create reports on a weekly or monthly basis so you know whether you are making a profit or losing money.

Keep your receipts! Every last one. Keep the receipt for this book, your supplies, utility bills, etc. Not only will you have an accurate record of where your money went, but you will also be able to write off a portion of your business expenses at tax time.

Review your books or accounting software periodically to see where you are spending the most money and find ways to lower those expenses.

WHERE TO FIND CUSTOMERS

Finding repeat customers should be your primary goal in the beginning. Here are some places to start looking:

- restaurants and juice bars
- health food stores and grocery stores
- farmers' markets
- individuals
- hold a cooking class for microgreens recipes
- hold a class to teach people how to grow microgreens for themselves
- be part of the community by participating in events, fairs, parades, fundraising, etc.
- advertise

Schedule time every week to search for new customers. At some point, as your business is growing, you may run out of time to do this. In order to expand your business to the next level, you may want to hire someone to take care of tasks that you don't personally need to attend to, such as cleaning.

FRESH SHEETS

You may have heard of the term 'fresh sheet'. A fresh sheet is like an

MY FARM NAME
www.myfarmname.com

~ fresh sheet ~
FOR DELIVERY ON: April 5, 2019

As the weather warms up we will be adding a variety of new crops to our regular rotation.
Thank you for your support.

~ products ~

micro sunflower	$14 lb
micro radish	$10 lb
micro pea shoots	$20 lb
salad mix	$20 lb
rhubarb	$3.50 lb

Call or text Name at:
123-456-7890

PLEASE ORDER BY APRIL 1
Call ahead for pick-up.

---

advertising flyer or a menu. It is a list of the products you have for sale for that week that you send out to customers. They can see what is available to order and plan their menus accordingly. It is a tool that can be used both in print form and digital. The key to using fresh sheets is consistency.

You can create a fresh sheet by making a table template in a word processor. It should contain your company name, contact information and the products available (this may change from week to week), quantities and prices. Be sure to include a time and date to order by, and the minimum order by price if you have one.

Locations with long travel distances, may have higher delivery costs.

It is wise to create a fresh sheet that is long and narrow so that it is easily readable if sent through text or messaging.

Leave your fresh sheet with potential customers and every week with your product when you make deliveries. Post it on your website, Facebook page and other social media. Send it out in a text message or email it to your customer database.

To the left is a sample of a fresh sheet. Feel free to include a short message or other foods you grow as well.

PITCHING YOUR PRODUCT

When you are starting out you will likely have to search out customers, as they will not have heard about you. One way to do that is by phoning and asking for a few minutes of their time to introduce your products.

Far more effective than phoning, however, is to speak with potential customers in person. Here is why:

- It is easier to dismiss someone over the phone than face to face. Arriving in person ensures that the customer has to engage with you.

- People like being able to put a face to the product.

- Meeting in person gives both parties more information in the way of body language and nonverbal cues.

- It allows for camaraderie, bonding and forming stronger relationships. It is preferable to do business with people we like.

- Face to face meetings are more memorable than phone calls.

- You are able to give away free samples and allow your product to speak for itself. Who doesn't like free stuff?

- The potential customer becomes more involved in the discussion instead of just listening to your pitch.

If you choose to pitch your products in person, make sure you arrive at a time that is most convenient for them. This is especially important when visiting restaurants because they are so busy. Do not arrive at their busiest times (meal times and preparation for mealtimes).

This should go without saying, but as much as you want your products to look top notch, you too should strive to present yourself well. While it isn't a suit and tie affair, it is a business meeting, and as such, you should conduct yourself professionally.

When you arrive, ask to speak to the head chef or the manager. Do not pitch your microgreens to the host or wait staff as they are not able to make purchases, but do let them know why you are there - you are a local microgreen farmer who would love to offer them your free samples.

Bring free, full-size samples, professionally and attractively packaged and properly labeled. Know your products inside and out. Also know what you are capable of producing, and what your turnaround time is.

Let the customer know that they can order right then and there. If they do not wish to order immediately, let them keep the free samples with the knowledge they can place an order at any time. Give them your business card and a fresh sheet. You ideally want to leave your samples and card with the head chef. Do not waste their time; chefs are busy people.

When a customer places their first order, set a schedule for future deliveries, right then. If you can set up payments to be made COD (cash on delivery) you will not have to collect payments between deliveries, or worse, show up repeatedly looking to collect on monies owed to you.

If the amount of product ordered is not worth your time and gas to deliver, say no. You may want to have a minimum order amount listed on your fresh sheet that reflects the distance you will be traveling. For example, a two-three tiered flat rate system based on miles from your operation would work well.

If a potential customer has not called you back within 3-4 business days, you should follow up, with the same person whom you left your samples with and spoke to. Again, arriving in person is much better than a phone call. Do not neglect to follow-through with a customer:

- It is a reminder for busy professionals who may have forgotten.

- It makes the customer feel 'special'. It allows you to demonstrate excellent customer service and lets them know they will not be left hanging.

- You are able to answer any questions they might have thought of since your last visit.

- Even if you do not get their business, they may have a lead or referral for you. (Do not be afraid to ask for a referral!)

If, after you have followed up with a customer and they are still unsure about placing an order, offer to come by regularly, every week at the same time, with a selection of products that they may purchase at that moment. Over time you will be able to identify their needs and always have the right products available.

DEVELOPING GOOD BUSINESS RELATIONSHIPS

It is important to develop good relationships with your customers. Aim to make it easier and more enjoyable for them to do business with you, than with your competition.

Make ordering simple. Keep packaging sizes to a minimum. Fewer choices in size are better. When presented with too many options it is easier to make no choice at all. 2-3 bag or clamshell sizes with a bin option for bigger orders will suffice.

Check in with your customers occasionally to offer new products (give free samples of them) and to get feedback. Be attentive and be consistent.

MARKETING

The following are marketing avenues that will help bring visibility to your business. Bear in mind that you do not need to utilize every one of these options. Often, word of mouth will drive your business more than advertising will. You may also find that some, particularly social media, can be time-consuming. Hunting down new customers or working your farm is often a better use of your time.

Website:

While not always necessary, a website is an excellent way to provide information to your customers. It is inexpensive and probably something you can create yourself. There are many website providers such as WordPress or Wix that offer easy drag and drop page templates which you can customize to fit your needs. A one-three page site should suffice.

When you register for a website you will have the option to purchase a domain name (a .com or the like). You will want a .com

over other extensions because .com is the one most people are used to seeing and using. Your web address should be simple and memorable and use standard spelling. Customers will not be able to find you if they are not sure how to spell your name.

All you need is a static page. Your website does not need a blog or constant updating (unless you enjoy doing those things and have time for them). Keep it clean and simple. Provide any necessary information as well as the best photos of your microgreens (or stock photos) that you can manage. Be sure it contains your contact information and product list. Consider adding a personal touch by way of the story of your business, the why and how it came about, and a photo of yourself with your product.

Facebook Page

Facebook is an immensely popular social media platform that can be easily utilized to reach potential customers. A Facebook Page can be used as an alternative to, or in conjunction with a website. And it only takes minutes to set up. The great thing about using this platform for business is the ease with which you can find customers. The downfall is that it is something people expect to be always up to date. If you choose this route, try to update every couple of days, with a quick message and/or photo and your weekly fresh sheet. You do not always have to update specifically about your business. You can reference online articles, other people's blog posts, recipes, microgreens facts, photos, etc. There are even apps for automatic updating.

Other Social Media

Consider creating videos for youTube, or posting a daily/weekly photo on Instagram. Many social media platforms are time-consuming and their audiences will have different expectations. If you choose to go this route, find one you enjoy doing that you can maintain regularly. Regular content is the key to a happy audience for most social media.

Facebook/Google Ads

The beauty of running ads on Facebook or Google is that you can be very specific in the audience you are targeting. You can narrow it down by region, age, interests, etc. You can target the health conscious, foodies, chefs, people who juice, etc. Plus you can set advertising spending limits. A word of caution... start slow and see what kind of results you get, then tweak your ads or your audience accordingly.

Vehicle Advertising

Take your advertising wherever you go by investing in a decal for your car or truck. Be sure it contains your contact information. Professional wraps look fantastic but are very expensive and probably not where you need to invest your money until you are making a regular income and can afford it. These are excellent ways to advertise for a one-time flat investment.

Always be prepared in case someone on the street asks about your microgreens business. Keep business cards, order forms, and a few fresh sheets in your vehicle at all times.

Merch

Merchandise is another great way to advertise your business. Put your company logo/website on your t-shirt, hoodie, bag, pen, phone case and key ring.

Farmers' Markets

Farmers' markets are not just a good place to sell your products. They are a way to get brand recognition and regular clients. Start small. Display your products vertically as it makes it seem that you have more product than is actually there. And make sure both your microgreens and booth look great! Use eye-catching signage. Mark prices clearly. Be friendly and talkative. This is part of the reason people love to come to farmers' markets, to connect with the people and have an experience. Consider handing out a small information card with your product extolling its nutritional benefits or a recipe idea.

OTHER WAYS TO GET NOTICED

- Get listed in the phonebook and local business directory.

- Write an article about microgreens for your local newspaper.

- Ask customers for a written review or quick quote that you can post on your website or social media.

- Leave your card at restaurants, post offices and anywhere you find a bulletin board.

- Network with other businesses that serve your market but who are not your direct competition, for example, specialty mushroom growers, artisan cheese makers, etc.

- Ask for referrals. You will not always get what you do not ask for.

BRANDING

Branding is a marketing concept used to develop an image for a product or company. It can be useful for getting customers to think about you in a particular way. And it will help your business to stand out from your competitors'. Your brand should be consistent wherever people find your company, whether online or in real life, or advertising. Here are some simple ways you can brand your business.

Adapt your brand to suit your target audience and what they care about and want. Ask yourself why people should care about your product. What are you offering them? Sum it up in the form of a mission statement and turn it into a tagline or slogan. From then on, your product, advertising, customer experience, and everything about your business should reflect your mission statement. For example, if your mission statement is about having the freshest quality microgreens, then freshness and quality will be your main focal point or theme in advertising, photos, customer service, and so on.

You want your brand to become not just recognizable but unmistakably yours and nobody else's. Think about well-known brands that you would recognize anywhere - McDonald's golden arches, Google's colorful type font. Walmart's slogan, "Save money. Live better." Or Nike's "Just do it." While your business may never become as big as these, branding will help provide a consistent customer experience that people will recognize and associate with your product.

Use consistent colors in your logo design and all communication - anything customers will see. This includes your website, social media, packaging, advertising, fresh sheets, and anything that will be reflected in their customer experience.

Another good way to brand your product is by maintaining your stock. For instance, if you have products on store shelves, make sure you check on them regularly and replenish them before they the unsold stock has begun to wilt or look bad. This will ensure that everyone who sees your product will know that it is synonymous with quality and freshness. It won't matter how good your logo or slogan is if your product looks terrible.

FINAL WORDS

In closing, I would like to say that the microgreens business as a whole is rapidly growing. People everywhere are discovering the amazing nutritional benefits and great taste of these tiny plants. Getting started now, while the industry is fresh, will enable you to make a name for yourself as competition grows. It is my wish to help others succeed by sharing my knowledge. I hope you were able to find some valuable ideas in this book.

ABOUT THE AUTHOR

Brandon Keady spent four years in college studying for his dream career. Several years after graduating and working in his chosen field, he realized he'd made a big mistake, but did not know how to make a change. He was burned out, had no life outside his job and little savings.

As a child, Brandon had loved his grandfather's farm. Unable to continue down the current path, Brandon quit his job, gave up everything and move three states away. He started growing microgreens in his spare bedroom until he saved up enough money to purchase his own farm. Today he is happily married with two children, three dogs, and 46 acres. He grows microgreens still, and a whole lot more.

And he thinks you can do it too!

www.ingramcontent.com/pod-product-compliance
Lightning Source LLC
Chambersburg PA
CBHW072234170526
45158CB00002BA/898